ƆOTmoney
GLOBAL CURRENCY RESERVE

Dot Money
The Global Currency Reserve
Questions & Answers

www.DotMoney.Cash
www.GlobalCurrencyReserve.com
www.DotMoneyBook.com

Eric Majors
www.EricMajors.com

Published by The Write For Right Project
www.WriteForRight.com

The Right For Right Project is an international, humanitarian
sustainable business project
of the publishing division of Xt Blue, Inc.
that is funded by a combination of donations, products sales and
the financial support of Xt Blue, Inc.
www.XtBlue.com
Please support the goals and work of the
Write For Right Project to help all artists in the world by buying
products sponsored by the Write For Right Project
and making donations to the Write For Right Project at
www.WriteForRight.com

CONTENTS

ACKNOWLEDGEMENTS

I acknowledge Derick Smith for his painstaking reviews of the material in this book and for creating the cover and artwork. I acknowledge the WriteForRight.Com project for the great work that they are doing as my publisher and in developing and publishing the works of other artists who come from challenging circumstances.

I acknowledge everyone in the global team of the Dot Money Project and the Global Currency Reserve (GCR) around the world who is helping to build the Dot Money and GCR systems in order to introduce the next age of global financial prosperity.

I acknowledge all of the people who are working around the world to help unlock the utility of money for the future of the human race by developing virtual and community currencies, and the technology that will facilitate a new world where money works for people rather than people working for money.

This is for you:
"We are not now that strength which in old days
Moved earth and heaven; that which we are, we are;
One equal temper of heroic hearts,
Made weak by time and fate, but strong in will
To strive, to seek, to find, and not to yield."

from the Alfred, Lord Tennyson poem Ulysses

Introduction

This book is designed to enable the reader to become familiar with the purposes and function of Dot Money and the Global Currency Reserve (GCR). The concept of Dot Money was established in the 2014 book "Dot Money" by Eric Majors (www.DotMoneyBook.com) and is currently being implemented throughout the world. This book is intended to provide a quick overview of the live business and implementation of Dot Money. For more information about the motivation, design and philosophy of Dot Money, it is necessary to read the original "Dot Money" book.

This book is written in the form of stand-alone questions and answers in order to help readers to quickly find the information that is most pertinent to them. In order to enable the most rapid acquisition and transfer of information much of the fundamental information is repeated in different forms in answer to different questions.

The book is written with some phrases bolded that can be read as highlights in order help speed up the reading process for those people who only want a high level overview.

Dot Money is a new kind of global community currency that incorporates the technology of virtual currencies and adds many new features that enable it to be used with or without computers or the internet.

The design and purpose of Dot Money is to introduce a new age of global economic prosperity and stability throughout the world and solve some of the most important problems facing the world today, including ending poverty.

Dot Money works in association with other currencies

and helps maintain the value of other currencies. Even though Dot Money can be used like money it is not designed to replace other existing currencies but provide a solution that does not currently exist in the global marketplace.

The Global Currency Reserve (GCR) is the administrator and primary market maker of Dot Money. The terms Dot Money and the Global Currency Reserve (GCR) are often used interchangeably.

Dot Money &
The Global Currency Reserve

What is Dot money?

Dot Money is a powerful private sector solution created by people around the world that will provide every peaceful person in the world with a monthly minimum living allowance, help to solve the problems of poverty, global economic instability, onerous taxation, the solvency of governments and inefficient welfare programs.

Dot Money is a real world implementation of the ideas put forth in the 2014 book "Dot Money" by Eric Majors (see www.DotMoneyBook.com) that are supported by many people around the world from all walks of life.

Dot Money is a new kind of virtual currency that builds on the successful features of BitCoin and community currencies like the Brixton Pound, but with an expanded set of capabilities and features that are specifically designed to help solve the major problems in the world. This includes helping to provide every peaceful person in the world with a minimum monthly income that enables them to maintain a minimum standard of living and strongly stimulate the economy. Dot Money is specifically designed to help reduce poverty, lower taxes, possibly eliminate the need for taxes, help support the value of all of the currencies of the world, create a stronger and more stable global economy, provide for security against manmade and natural disasters and provide for the pursuit of health, happiness and the general welfare of all of the peaceful people of the world. The technologies and regulations associated with Dot Money are being created in association with and to serve individuals, businesses, banks and governments of all sizes around the world.

The primary administrator of Dot Money is the Global Currency Reserve (GCR) and thus the names Dot Money and the Global Currency Reserve (GCR) can be used interchangeably.

The plan of the Global Currency Reserve and Dot Money is to provide every peaceful person in the world with the minimum income necessary to pay for food and lodging, and to help cover the costs of medical care, education, transportation and communication, **without taking anything from business, government or anyone else, without re-distribution of wealth or trying to make everyone financially equal. Dot Money does not take from the rich to give to the poor or derive its resources from government budgets to achieve its goals.** Dot Money's unique business plan does not take any money from anyone in order to pay for these expenses but rather Dot Money adds a critical safety net for everyone in the world and their respective governments. Dot Money allows people who have the good fortune of financial success to keep their earnings while supporting those who find themselves in financial distress for any reason. Healthy, well funded consumers make for a healthy economy, thereby increasing the health and welfare of the entire population of the world.

The power of Dot Money resides in the people who support Dot Money, who convert to Dot Money and insist on transacting in Dot Money in order to promote the values and goals of Dot Money to create a better more stable world.

What are the units of Dot money?

One unit of Dot Money is called one "Dot" which is not created and backed by debt but is created and valued based on the value of the exchange of tangible capital and human willpower and, more specifically, by the desire to use Dot Money in exchange for goods and services, in support of a stable global economy, and the health and welfare of all people in the world. **You can think of the value of one Dot as equal to the purchasing power of one U.S. Dollar on December 1st, 2014.**

Dot Money's unique features and business plan allows for the exchange of Dot Money to and from various currencies around the world within tightly fixed exchange rates determined by the **Global Currency Reserve (GCR)**, which is the primary global liquidity provider for Dot Money. The GCR works with governments, businesses and banks in their respective territories in order to help establish the boundaries for fixed rates of exchange between Dot Money and every GCR eligible currency, that will effectively be supported by the market making activity of the GCR, using the tool of Dot Money. The GCR uses Dot Money as the primary record of exchange, whose value is sustained and defended by the GCR within specific rates of exchange.

The goals of the GCR and its use of Dot Money are beneficial and help to create global economic and monetary stability for all countries of the world. By buying and selling Dot Money, and making a market for Dot Money at fixed rates of exchange with other GCR eligible currencies, the GCR will also indirectly help make and maintain a market in each of the currencies of the countries that allow Dot Money to be traded or used by the citizens of their countries. These predetermined ranges of exchange rates between Dot Money and GCR eligible currencies will remain fixed and may only be adjusted by the GCR on a case by case basis, only if completely necessary, or only if some extraordinary need arises.

As an example it is the intention of the GCR to fix the rate of one Dot to within in plus or minus a fixed percent to one U.S. Dollar as it is valued based on the buying power of the U.S. Dollar as of December 1, 2014. The rates of exchange of other GCR eligible currencies will either be based on the prevailing rates as of December 1, 2014, or adjusted after research and negotiations between the GCR and each country that allows the use of Dot Money (GCR eligible countries and currencies). The exact rates of exchange (and the upper and lower boundaries) between Dot Money and GCR eligible currencies will be announced and updated as needed upon the official launch of Dot Money.

What is the list of GCR eligible countries supported by Dot Money?

Dot Money and the Global Currency Reserve (GCR) are committed to work with, and assist, governments, businesses and individuals around the world for the benefit of mankind. However, Dot Money and the Global Currency Reserve will not operate in any jurisdiction where the local laws prohibit the business of Dot Money or make the use of Dot Money too onerous.

GCR eligible currencies that can be used to purchase Dot Money, and into which Dot Money can be converted, are those currencies of governments who participate, cooperate with, or support the Global Currency Reserve (GCR) and the Dot Money program. Dot Money may also be eligible for use with currencies of governments that do not oppose the GCR and the use of Dot Money, and that do not create any regulatory infrastructures that makes the operation of Dot Money within their countries too onerous.

Some of the countries whose currencies are expected to be eligible for exchange with Dot Money include the U.S.A, Canada, Mexico, member countries of the EU, South Africa, Hong Kong and China. The GCR is also in preliminary negotiations with other countries, such as Russia, to determine the viability of their participation as a GCR eligible currency to be supported by the market making activities of the GCR.

A complete list countries whose currencies are currently under consideration for GCR eligibility and exchange with Dot Money can be found on the website of the Global Currency Reserves (GCR). www.GlobalCurrencyReserve.com.

Where can I spend Dot Money?

Dot Money can be spent where people and businesses accept Dot Money. BitCoin is a good example to follow because it is now being accepted by more and more major vendors, paving the way for Dot Money to also be accepted by these same vendors. Dot Money is also unrolling a unique method of point of sale payment system that will enable people to make retail purchases at store without having to carry any paper money, checks or credit cards. Users will simply be able to provide certain numbers to the cashier and the payment will be effectuated. In this way Dot Money will be reduce financial crime and fraud and will be safer to use than carrying other forms of payment.

As an incentive for vendors to use Dot Money, Dot Money administrators will eventually pay the sales tax and vat on Dot Money transactions.

In the early stages the Global Currency Reserve's (GCR) (third party) payment gateway will further enable vendors to stipulate that any or all portions of payments made to them in Dot Money are be instantly converted into their native currencies in order to complete the sale. In this way vendors can accept Dot Money at no risk. The goal is that Dot Money will eventually be accepted as a means to purchase anything from anyone and the facility of the GCR's Dot Money payment gateway provides for instantaneous payments to vendors in their native currencies, which should enable any organization to start collecting payments in Dot Money as soon as it is released. When Dot Money is released and available for use, Dot Money will be convertible back to any other currency at any time by users at fixed rates. Thus in this way it should technically be able to be used by anyone to purchase anything, even in the initial stages, by converting to other currencies. However, it should be noted that supporters of the concepts of Dot Money should try to use Dot Money in transactions and ask vendors that do not already accept Dot Money as payment to start accepting it in order to help ensure the success of the goals and use of Dot Money.

Why is Dot Money needed?

We have come to a point in human history where our experience indicates that it is not gold or silver or any commodities, other than human will, that establishes the value of any currency. **Without people agreeing on the value of money, money itself has no value or use. Human beings are the only true origins of the concept of value in the world and it is now possible to create a global medium of exchange that leverages this obvious fact.** Whether someone is gainfully employed,or for whatever reasons they are unemployed and in need of financial assistance, every person in the world who spends money is a critical contributor to the economy. Critical contributors to the economy include people who do not work at all but purchase food, clothing and other products that are needed to sustain their lives.

Business owners are in need of paying customers whether these customers are rich or poor. The less money that is available for people to spend the less healthy and stable the economy. Current implementations of debt based global money systems are limited, inequitable, and incomplete, since the problems of the expansion of the population were not adequately contemplated during their implementation.

As a result of the original design of modern debt based monetary systems that are still in use today most major currencies now exist in the same way as any other scarce commodity, the supply of which is simply inadequate to service the needs of the users of money. Thus the entire global economy is experiencing unintended and damaging waves of economic instability. The only remedy to overcome the problems with the current global monetary supply is through the creation of an additional liquidity component to the global monetary system that allows for the creation of money that does not need to be paid back. The monetary supply must be allowed to expand and contract based on the number of users of the money itself (rather than demand computed by spending data) and the valued of money must be based primarily on agreement between users in order for it to retain its value and usefulness.

This model of liquidity and value is based on the global stock markets which already establishes and sustains the value of relevant businesses whose shares are issued and traded in the financial markets.

What is wrong with our current global economy and monetary system?

As the population of the world expands and contracts, so should the supply of money. The current debt based money systems simply create a new commodity whose scarcity increases as the global population increases. Creating temporary money that must be paid back from lending is not enough to solve the critical problems of lack of money in today's global economy.

To illustrate the problem consider that, if every person in the world were gainfully employed at a subsistence level minimum wage, even if you had access to all existing currency in circulation from all of the nations of the world, mathematically there would not be enough currency in existence in the world today to pay each person and allow them to save as required to survive for their retirement.

As a result of this lack of supply of money that does not need to be repaid, everyone in the world is experiencing the booms and busts associated with ebbs and flows of what little money is concentrated by the wealthy into the next financial bubble. Because of the lack of money in supply the entire practice of investing using the global financial markets has come to resemble little more than a regulated Ponzi scheme. In a Ponzi scheme, just as in our financial markets today, the first investors into the next bubble (investment concentration) are financially rewarded as they exit the market and take their profits, while everyone else is left with little or no money and little or no value.

Most people believe that too much money in circulation creates inflation. However, the specific type of inflation that is being experienced today is being caused by the scarcity of money available to purchase goods and services. When less buying and selling is taking place, providers of goods and services must increase the prices in order to sustain themselves in a marketplace with fewer sales. When the economy recovers most businesses do not lower their prices as sales of their products increase. As a result of the lack of supply of money itself the entire world is experiencing the bankruptcy of governments, banks, businesses and individuals. When people do not have enough money to repay loans, banks and financial institutions fail, creating more uncertainty and less financial security. As a result of the lack of supply of money and the associated inflation there is nothing that can be purchased with currency that will retain its value without risks of investment speculation. The question that many people are asking is what can be done to protect the value of their money in today's unpredictable global economy? Investing and speculating into other commodities does not provide any certainty.

Why can't the government solve the problem themselves?

There are ways that the major governments of the world can correct these problems themselves as described in www.DotMoneyBook.com. However, the entire system to remedy the global economic instability is already being implemented much more quickly and efficiently by a business that is not controlled by the political interests of any single country or slowed by its bureaucracy. That business is called the Global Currency Reserve (GCR) and its creation and use of Dot Money does not require any government to take any action at all other than to allow Dot Money and its business to be conducted without onerous limitations.

A privately held international organization that works with people all over the world has more freedom to take action that is beneficial to all of the people of the world, rather than only those people in certain countries. A privately held international business concern created by ordinary people, like Dot Money, that abides within the laws and cooperates with governments, but is not under the direct control of any of the governments or other entities that have brought us to the brink of our current global economic instability, will be able to gain the support of the people that is serves (the customers, and users of Dot Money).

Dot Money and the solutions for governments, businesses and people will not be available in countries where use of Dot Money is not legal or is made too onerous by local rules and regulations. As long as the majority of the people in the world are happy with the performance of Dot Money, and its administration by the Global Currency Reserve (GCR), then Dot Money will persist, but if the management of Dot Money does something that angers the users then the system will cease to be relevant and it will end because people will stop using it.

Is Dot Money a replacement for existing currencies?

As in many cases throughout history a combination of solutions from the private sector working with governments are the best cure for problems. Dot Money is this private and public cooperative solution that can provide value to all global currencies and liquidity for governments and their citizens who work with Dot Money. Dot Money has similar features to money and can be obtained and spent just like money, but **Dot Money has many more features than ordinary money. Dot Money is not a replacement for the currencies of the world but can be used to support the value of all currencies of the world. Dot Money does not have a debt based creation system that is needed and provided by local governments and their respective banks, in their own currencies, for lending and temporary expansion of the global money supply. Dot Money provides a critical tool for the creation of money in a way that will be necessary for the governments of the world to solve the problems facing them.** Like it or not the fundamental solutions espoused by Dot Money will need to be employed in one way or another, either by governments or in the private sector. As of today Dot Money has laid the ground work and established the necessary global relationships and cooperation with business and government to ensure that Dot Money is in the best position to be the primary public, private and commercial cooperative solution around the world.

Dot Money and the Global Currency Reserve (GCR) works with governments and their respective banks to help to ensure that all loans taken out by ordinary people and businesses can get paid back. Dot Money works with governments and their respective banks to help ensure that they have sufficient capitalization to maintain their functions and lower or completely eliminate their national debts. **Dot Money works to strengthen the currencies of the world by**

making a market for all currencies within tightly fixed ranges of the rates of exchange. Dot Money does not tamper with the existing systems of global currency exchange, but Dot Money adds to the world a valuable and necessary tool that provides stability for all of the currencies of the world that are exchangeable with Dot Money. Currently the list of GCR eligible currencies that can be converted to and from Dot Money includes most of the major currencies of the world. **The most current list of GCR eligible currencies can be found on the website www.GlobalCurrencyReserve.com.**

What laws, rules and regulations are applicable to Dot Money?

The Dots of Dot Money exist just like any other object after it is created by the Global Currency Reserve (GCR). Because of the global distributed information network (and backup, human based, network of Dot Masters) that supports Dot Money, **once Dot Money is created it cannot be destroyed or controlled by any third parties (including governments) outside of the original functionality for which it was designed**. Dot Money is a virtual record, and there is nothing to be regulated or controlled about a virtual record itself. In other words, the functions and utility of the units of Dot Money are not subject to any regulations or laws as they as they simply exist as a set of rules and records.

That is not to say that the Global Currency Reserve (GCR) is not subject to the laws and regulations in the countries in which it conducts business. Just like any other substance or process, the end **use of Dot Money is subject to the rules, regulations and laws of the various jurisdictions where the use of Dot Money is permitted. Most laws and regulations regarding Dot Money and other virtual currencies have to do with the ancillary businesses and transactions that facilitate Dot Money**, such as the regulation of exchanges that exchange currencies including Dot Money. Other businesses that use Dot Money that are also usually subject to laws, regulations and registrations by their local governments include businesses which attempt to store value or hold Dot Money on account for others or facilitate buying and selling with Dot Money.

In most cases the Global Currency Reserve (GCR) does not conduct any regulated businesses within countries other than the buying and selling of Dot Money for its own account. As such, the Global Currency Reserve (GCR) and Dot Money are not directly involved with any regulated businesses. **The Global Currency Reserve (GCR) and Dot Money works with governments and their law enforcement to help prevent crime and the abuse of Dot Money** that may result in the breaking of laws in any countries. **Any currency in the world, including existing paper money and coinage, can be used by criminals for illegal purposes**, however, the GCR and **Dot Money** do their best to help law enforcement and Dot Money does provide **extended facilities for crime detection, prevention and restitution**.

The GCR takes every precaution possible to ensure that any **businesses that work directly with the GCR,** such as currency exchanges, banks and other financial institutions or transaction facilitators, **are all properly registered and in good standing as may be required by the rules and laws of their respective countries.**

What are some ways that Dot Money helps to stabilize the global economy?

Dot Money provides governments, businesses and ordinary human beings with a way out of the global economic instability and into financial prosperity for the good of all.

From a technical standpoint **Dot Money is a tool that seeks to establish and maintains its value through a combination of ordinary market making and cooperation from people around the world, who willingly choose to use and promote the use of Dot Money** so that the Dot Money system can achieve its humanitarian goals, which are beneficial to governments, business and everyone in the world. When Dot Money is strong so are all of the other currencies of the world which are exchangeable for Dot Money within specific price ranges.

What is the difference between Dot Money and other money?

Dot Money is the next logical evolutionary step in the history of human economics. **Dot Money has many more features than ordinary currencies and commodities.** Dot Money borrows the successful technology and lessons learned from the history of the first bankers of recorded time, to modern day governments, their national or reserve banks, and modern day virtual and community currencies. More information about the history of money and Dot Money can be found in the book "Dot Money" at www.DotMoneyBook.Com.

Technically speaking Dot Money is not money at all even though it has many of the same features and uses as money and virtual currencies. At its core **Dot Money is an advanced record of exchange for value that can be bought and sold around the world. The value of Dot Money is established by agreement of its users (subscribers)** and as such the value of Dot Money is not subjected to the same risks of devaluation as the rest of the global currencies. Therefore **Dot Money preserves its value for those who use it regardless of any economic conditions around the globe.** Unlike all other existing currencies, the **users of Dot Money agree to exchange it only within specific price ranges with respect to other global currencies.** In addition, unlike existing popular virtual currencies such as BitCoin, **Dot Money has an administrative body that not only helps to preserve the value of Dot Money but will also be able to reduce crime and fraud associated with the use of Dot Money.**

In addition, **Dot Money will implement a human based backup network of ordinary people, called Dot Masters,** that will be hired and paid by the GCR **to enable Dot Money to be used even in the most remote locations, with or without computers or any other modern technology or electricity.** Thus **if modern internet, electrical or software systems are ever compromised Dot Money will continue to persist** and retain its usefulness, value and systemic integrity. In the short term **Dot Money provides a place where people can truly record, store and retain the value of their money over time** that is unlike any other existing methods of capital preservation that are currently promoted.

Is Dot Money an investment product?

Dot Money is NOT an investment product, or security of any kind, but a new tool that can be used to retain value over a period of time. The value of Dot Money is not backed by anything other than the will of the people who use and benefit from the system. **Buying, using and selling Dot Money is the same thing as buying and selling any other object that does not come with any guarantees of anything other than ownership.** The Global Currency Reserve (GCR) is a privately held organization that purports to be a large market maker in Dot Money and the administrator of Dot Money, but the purchase of Dot Money does not guarantee the purchaser of anything other than ownership of Dot Money. **The success of Dot Money depends on the desire and the will of its users, supporting governments, and secondary and retail markets to use Dot Money.**

Specifically, Dot Money is a record that people can buy, sell and trade. There is no need for Dot Money to be held on account or stored in any bank or financial institution and, as such, **users of Dot Money are not at risk of having their Dot Money disappear or become inaccessible as the result of the failure of any governments, banks, businesses or institutions.** Even though Dot Money is not held in a **bank, Dot Money is not designed or capable of replacing existing global currencies. Dot Money is used in cooperation with banks, and Dot Money creates highly profitable business for banks using their own native currencies** and conducting their business as usual. Thus, **Dot Money is not a threat to any financial institutions or investment businesses**, but rather helps to create stability and liquidity for these businesses and their clients.

All other forms of capital preservation being promoted are essentially only forms of speculation in property, commodities or investment products. Dot Money is not an investment product at all but a group of users whose membership is based on an agreement between its members to honor the exchange of value. **Thus Dot Money provides a unique tool for people to retain value that is unlike anything else that currently exists in the financial marketplace.**

Dot Money is an agreement between its users that says that no matter how unpredictably the global economy behaves, and no matter what manmade or natural disasters may occur, Dot Money will still retain its value. Dot Money is not mandated by any government or organization, but is an agreement amongst all Dot Money users and subscribers.

What does Dot Money have to do with barter trade?

Dot Money is an agreement amongst its users to record exchanges of value that more closely

resembles barter trade than the exchange of money. Users of Dot Money agree to hold the value of Dot Money within a specific rate of exchange range so that holdings in Dot Money will still have at least the same buying power at a future point in time as it did when it was purchased in the first place. **Ultimately the Global Currency Reserve (GCR) intends to use its infrastructure to help facilitate pure barter transactions around the globe.**

Why is there a need for people to receive a monthly living allowance?

The prosperity of the global economy depends on consumers with money to spend. Dot Money does not make people rich but it does provide people with an increased opportunity to pursue their peaceful passions, get an education or work with pride in lower paying jobs while genuinely increasing their overall standards of living. Dot Money recognizes that the accumulation of financial wealth is not the highest priority for every person in the world and that life should be more than just trying to make ends meet. **Dot Money solves the age-old problem of ensuring that every person in the world is adequately compensated for the segregation of the property in the world** that would otherwise be accessible to everyone equally if it were not for the forms of government upon which we have all become dependent. Governments primarily regulate the use and distribution of property. **Since governments are an essential part of modern life, Dot Money compensates all people for the use of all property that has been taken under the control of their respective governments.** This compensation is necessary because people are unable to "live off the land" that is all either privately owned or owned by the government.

It is this age-old "taking away" of something by

government to give to someone else, which has never before been properly contemplated or addressed. **It is this fundamental failure of the structure of all governments in history that they have failed to provide compensation for what has been taken from both the rich and the poor. This "taking away" without giving enough in return has consistently created resentment by people against governments** throughout history. The effects of this collective consciousness of resentment amongst the citizens of the world are now building into an angry, and often chaotic, expression by people who are using global communications technology to join forces. **These modern sentiments and expressions of unrest are adding to the destabilization of the global economy and rule of law as people are losing faith in governments, banks, businesses and each other.**

This modern day unrest manifests itself in the arguments between groups who appear to be ideologically opposed to each other but are actually both articulating their displeasure about what has been taken from all of them. The fundamental arguments of the most popular groups today are all in agreement that something is missing or being taken away from them in terms of finance and economics. **A list of the popular groups that appear to be arguing over scarcity of financial resources in the United States includes, Occupy Wall Street, small and large business owners, the Republican Party, the Tea Party, the Democratic Party, Libertarians and Independents (the left, the right and the middle).** People from all political parties and philosophies all appear to acknowledge that there is not enough money to go around and that the current offerings of government services including taxation and welfare are dysfunctional, inadequate and unsustainable. **Dot Money provides a non-partisan solution that satisfies the needs of all concerned groups without taking anything away from anyone.**

The current construct of debt based money systems, financial welfare systems, and virtually all other services offered by governments of the world, depend on heavy and unpopular taxation and redistribution of the wealth. Almost all people, regardless of their political or other affiliations, agree that the current solutions being provided by government are inadequate and financially unsustainable. **One primary reason why the current government systems are unsustainable is due to the rates of change of the population of the world and lack of monetary supply. The solution is to create another avenue of liquidity based on agreed value, which is what Dot Money seeks to provide for governments, businesses and individuals.**

In addition to providing real and substantial benefits for individuals throughout the world, **Dot Money also works with business and governments to help them to lower costs and reduce crime.**

Why not just go back to the gold standard?

Money and its value exist as an abstraction that can be recorded and virtualized. Going back to the gold standard would be more useful in helping to stabilize the global monetary systems and economies than continuing with the status quo. However **there are more efficient ways to stabilize global currencies than using the gold standard, i.e.** using the latest technology and consciousness about money that exists today.

Using price controls to fix the value of money to any specific commodity, as in **the gold standard, is redundant and unnecessary, and will ultimately introduce other problems**. **Some of these problems include the** random nature in the occurrence of the commodity which creates **random and uncontrollable concentrations of wealth and financial lack** between various countries in the world. Another practical problem with using a gold standard is the possibility that, **over the course of time, the actual demand and perceived value of gold may required further adjustments and redesign of such commodity**

based money systems if the nature of our relationship to gold should ever change.

Since all we are really concerned about is the relationship between values of global exchange (for example maintaining the value of currencies in relation to each other) **it is much easier to virtualize and record transactions of value that are independent of any particular commodity. Value records of exchange, such as Dot Money have many more features than paper money or gold, and are far more flexible and able to adapt to any unforeseen future technological and economic needs of human users.**

Price controls through the act of law that attach value to specific commodities can always be done by governments in any case without having to "base" their currency to any specific commodities.

Why will governments, businesses and people want Dot Money?

Dot Money helps governments by subsidizing or, in some cases, completely eliminating the costs to governments associated with the provision of welfare programs, and the associated burdens on tax payers. Dot Money does not take money away from anyone else to assist governments and provide a basic minimum income for everyone in the world.

Dot Money helps businesses by improving the financial disposition of their employees, reducing taxes and allowing businesses, their owners and employees to keep more of their earnings.

Dot Money helps individuals by providing them with a minimum monthly living allowance that enables them to focus on what is truly of importance to them. The provision of a basic monthly income enables people to live lives where they are not slaves to money, but where money works for them and there is at least enough money at all times to help people to discover and unlock their true individual potential.

The foundational elements of Dot Money are already being successfully used throughout the world thus ensuring the success of the Dot Money movement. Dot Money is a currency for good that represents the next evolutionary step of global economics. Dot Money is evolution being realized today.

What is the difference between Dot Money and Community Currency?

Investopedia.com defines community currency as *"A form of paper scrip issued at the county, town or community level for use at local participating businesses. The theory behind community currencies is to encourage spending at local businesses as opposed to chain or 'big box' stores."*

There are many examples of successful community currencies including, the Brixton Pound (**http://brixtonpound.org**). **Dot Money is also a community currency but on a global scale for a global marketplace.** Dot Money does make use of some of the same software technology that is currently being successfully used by the administrators of community currencies.

The community that is served by Dot Money is all peace loving people around the world who want to end global economic instability and who are tired of living under the threat of currency devaluations and unstable financial and investment markets. The community served by Dot Money are those people who want **to reduce or eliminate taxes, help to keep their governments, businesses, citizens and economies from going bankrupt, and who want to provide a basic subsistence level income for all other peace loving people in the world in order to**

end poverty and improve the overall health and welfare of all people around the world. This is inclusive of the rich and poor and doesn't take anything from anyone. The community of those served by Dot Money are **those who are in favor of implementing a new age of monetary policy and global financial prosperity, where people do not exist as slaves to money but where money is a tool that is of benefit to all people, businesses and government, without any burdens being placed on any government, business or group of people**. Some refer to Dot Money as "Star Trek" economics, but Dot Money is really a long overdue medium for exchange.

The implementation of Dot Money by the Global Currency Reserve (GCR) utilizes a distributed network of software systems and human backup systems known as "Dot Masters" to record the ownership and the exchange of Dot Money throughout the world. Thus Dot Money is constructed with a backup system comprised of a human network and **Dot Money does not require the use of a bank or other financial institution to hold Dot Money on account. Technically Dot Money is not a currency at all but rather Dot Money exists as a record of transfer of value for value.** However, in order to help facilitate the transactions between Dot Money and the native currencies of GCR eligible countries, Dot Money works with Banks around the world to ensure and promote seamless exchange between Dot Money and the native currencies of businesses who accept and transact in Dot Money. The process where the GCR makes a market in Dot Money is extraordinarily simple. It creates a very low risk, highly profitable and lucrative stream of profits for banks and financial institutions who allow Dot Money to hold accounts in their institutions, and conduct the buying and selling of Dot Money for the account of GCR in the native currencies of GCR partner banks.

Even though there are strong relationships between Dot Money and banks around the world, neither Dot Money nor the Global Currency Reserve

(GCR) are banks. As such the GCR and Dot Money do not engage in the businesses of providing financial accounts, investment advice, or selling investment products. Dot Money and GCR also do not act as a currency exchange, directly facilitate financial transactions, or provide financial transactions services between any third parties. **The Global Currency Reserve(GCR) is a private institution that buys and sells certain currencies for its own account including Dot Money and other currencies that GCR deems as "eligible" for exchange with Dot Money.** Because the rates of exchange between Dot Money and the other GCR eligible currencies of the world are fixed, GCR also directly makes a market in global currencies, helping to sustain them at or above specific rates in order to help stabilize the global economy for the good of humanity. The GCR is also the administrative body that promotes and governs the issuance, operations and use of Dot Money as a means to help improve the world.

What is the difference between Dot Money and BitCoin?

The Dot Money record keeping software system uses technology derived from the BitCoin source code to create a new standalone, global record keeping system that is substantially different from BitCoin. **Dot Money is specifically designed to help achieve certain constructive goals that were never contemplated by the designers of BitCoin.** You can think of Dot Money as a substantially enhanced version of BitCoin, with features that make Dot Money distinct.

Although Dot Money and BitCoin can be received, stored and spent in a similar fashion, there are many substantial differences between Dot Money and BitCoin since Dot Money was designed for entirely different purposes. Some of the fundamental differences between BitCoin and Dot Money are as follows:

1. *Creation and acquisition*. In order to obtain newly created BitCoins, an increasingly expensive and elaborate electronic mining effort must be undertaken that can be as costly and unpredictable as mining for gold. The creation of a new "Dot" (unit of Dot Money) is under the exclusive control of the Global Currency Reserve (GCR), which makes the primary market in GCR eligible global currencies including Dot Money.

2. *Rates of exchange*. The BitCoin virtual currency exists as any other commodity whose rate of exchange fluctuates from day to day. Thus the value of money used to purchase BitCoins is subjected to the fluctuations in the value of BitCoin in the global currency markets and the value of BitCoin has proven to be as highly volatile and unpredictable as any other commodity. One of the purposes of Dot Money is to provide a safe haven from rapid devaluations of global currencies for users of Dot Money. Thus, rates of exchange of Dot Money is intentionally fixed by the Global Currency Reserve (GCR), who makes a market in Dot Money in order to help sustain the value of Dot Money and, in so doing, indirectly makes a market which defends the value of all of the currencies of world that are GCR eligible (most major currencies). The latest rates of exchange between Dot Money and the rest of the global currencies of the world can be found on the website of the GCR at www.GlobalCurrencyReserve.com.

3. *Talk Back Capability*. **Users** of Dot Money are able to cast their opinions to the Global Currency Reserve (GCR) in all matters concerning the evolution and regulation of Dot Money through a real time voting process that is facilitated through the possession of Dot Money.

4. *Use with or without the internet*. Presently use of BitCoins without internet or computer access can be a difficult proposition and almost impossible at ordinary points of sale, such as supermarkets. Dot Money comes with a technology that enables users to spend money at any location, without the need to carry any paper money, plastic cards, cell phones or have any internet access. In this way using Dot Money is safer than other forms of money and includes multiple layers of protection to protect against financial fraud and theft.

5. *Biological record & backup methods.* BitCoins and their associated records of ownership and transactions exist in software that resides online on a distributed network of computers around the world. Dot Money records exist in the form of electronic distributed network records, however Dot Money is also working to institute a unique and powerful capability where Dot Money records will also be stored manually by human beings in duplicate sets of physical records that are independent of electronic, internet or communications technology. Dot Money is being designed with a dual compliment of human Dot Masters so that Dot Money can be used with or without computers, cell phones or any internet technology even in the most remote regions of the world. Using this new system of human Dot Masters, Dot Money will be secure from electronic and internet technology failures, hackers and less susceptible to fraud.

6. *Time Released Versions*. Dot Money has a feature known as "Time Release" that does not exist in BitCoin. Dot Money can be issued by the Global Currency Reserve (GCR) encoded so that the recipient of the Dot Money cannot spend the Dot Money until a specified date. Time release capability allows people to become liquidity providers and gain a profit for purchasing and holding Dot Money for a specified period of time.

7. *User Defined Time Release and Conditional Release*. Users who spend Dot Money may also program their own time releases into the Dot Money that they exchange. When a mutual agreement exists between the remitter and recipient then, upon acceptance of the time released Dot Money by the recipient, the recipient can only spend the Dot Money after a specified date or after a specified condition exists (escrow, buyer protection).

8. *An opportunity to make money (Liquidity Providers)*. There are some people who "got in on the ground floor of BitCoin" and made small fortunes when the price and liquidity of BitCoin became established around the world. Today those who wish to make money from BitCoins in some way must become speculators, who deal in BitCoins hoping that the values of BitCoins in the open market will work to their advantage. Alternatively some people still try to obtain new BitCoins through the ever increasingly expensive and unpredictable process of electronic mining of BitCoins. Those who mine for BitCoins are gambling that the extraordinary costs in time and money associated with the mining activity will result in enough BitCoins, that are sufficiently valued by the global markets, to cover the costs of the mining operations and also provide for some profit.

Because the design and goals of Dot Money are different from BitCoins, and because Dot Money is designed to be exchanged within fixed rates with respect to all other Global Currency Reserve (GCR) eligible currencies, there are more predicable ways for those who desire to make a profit in relation to working with Dot Money, with substantial benefits and easy to understand risks.

For more information about how to get involved in the launch of Dot Money, look for the opportunity to become a crowdfunder by making a non-tax-deductable donation to the Dot Money Project by going to the GCR website www.DotMoney.Cash.

Anyone can work with the Global Currency Reserve (GCR) after the launch of Dot Money by becoming a "liquidity provider" for the GCR. Liquidity providers purchase Time Released (or Time Coded) Dot Money that can only be spent at some time after purchase from the GCR. Time Coded Dot Money is sold at discounted rates that are established by the GCR and depending on the length of time that the purchaser agrees to hold the Dot

Money will be Time Coded for spending by the owner at a future date. By purchasing and holding Time Coded Dot Money Liquidity Providers help the GCR to make a market in Dot Money and sustain the value of Dot Money and all GCR eligible currencies around the world. Anyone or any organization can be a Liquidity Provider for Dot Money and the GCR, and Liquidity Providers help to stabilize the entire global economy and fulfill the beneficial goals of the Dot Money project.

Liquidity providers help provide the Global Currency Reserve (GCR) with additional capital to make and sustain a market in Dot Money and the rest of the GCR eligible currencies around the world. Because the transaction of the purchase of Time Released Dot Money is complete at the point of sale the purchaser becomes the owner of record to the Time Released Dot Money, and there is no need for any financial institution to hold any money or keep track of the Dot Money on behalf of the purchaser. Thus buying Time Release Dot Money provides the purchaser with protection from the default or collapse of any financial institution.

Currently Dot Money and the Global Currency Reserve (GCR) are not yet selling Dot Money or Time Coded Dot Money. All forms of Dot Money will only be available to be purchased using GCR eligible currencies after the launch date of Dot Money. The tentative launch date of Dot Money can be found on the Dot Money website (www.DotMoney.Cash). The only way to "get in on the ground floor" from the launch of Dot Money is to participate as a crowd-funder of the GCR and the Dot Money Project. For more information on how to become a crowd-funder of Dot Money and show your support for Dot Money go to www.DotMoney.Cash.

9. *Works with law enforcement.* Unfortunately, BitCoin has gained a reputation as a tool that can enable anonymous exchanges of money that are difficult to trace and can be abused by criminals. While transactions in Dot Money are performed in a similar manner as BitCoins, there is no central administrator for the creation of BitCoins like there is for Dot Money. As a result, there are other ways that the Global Currency Reserve (GCR) and Dot Money can assist law enforcement around the world to help apprehend criminals. The Global Currency Reserve (GCR) and Dot Money are created to help governments, as well as individuals, and as such the GCR will work with governments and their respective law enforcement to help fight fraud and crime whenever possible.

9. Dot Money *Business Model, Incentives & Taxation*. Dot Money was created with different goals than BitCoin. As such there will always be a way for individuals and institutions to make money by assisting the GCR and Dot Money by being Liquidity Providers for the GCR and purchasing time released Dot Money at discounts. In addition, Dot Money's business model includes sustainable incentives to encourage people and businesses to hold their assets in Dot Money as long as possible, while still providing major opportunities for banks and financial institutions to improve their profits and benefit financially by working with GCR in their native currencies. People holding Dot Money actually creates profits for their local banks who work with the GCR.

One of the initial and additional planned incentives that are designed to encourage people to use Dot Money include Dot Money paying the sales taxes of certain transactions that are conducted in Dot Money.

Another major incentive to use and transact in Dot Money will be the offering for peaceful subscribers to receive a monthly living stipend to help provide a minimum amount of money to each person to pay for food, rent, medical, education, transportation and communication.

Banks stand to benefit from keeping accounts for GCR that are used by GCR to buy and sell Dot Money throughout the world. GCR will typically only borrow against assets to buy and sell Dot Money. In addition, because neither the GCR nor Dot Money are commercial lending institutions, banks that hold accounts for Dot Money will have the luxury of large deposits on hand to help facilitate their lending businesses.

Governments and businesses alike will benefit from substantially lower expenditures associated with human resources including reduced expenditures associated with the provision of health and welfare services. This is because of Dot Money's program to ultimately provide every peaceful person in the world with a minimum monthly stipend. Governments will also benefit from Dot Money by having the value of their own currencies bolstered by the market making performed by the GCR in sustaining Dot Money. Governments that actually accept Dot Money as payment for taxes and government services will be able to substantially reduce and, in some cases, completely eliminate their national debts over time. It will be at the discretion of the management of GCR to provide specialized liquidity services to the reserve banks and governments of the GCR eligible countries.

What are the ideals and goals of Dot Money?

Dot Money seeks to provide the following solutions for users of Dot Money and help to make these solutions possible by promoting and incentivizing the use of Dot Money around the world:

1. End Poverty by providing every person in the world who becomes a subscriber with monthly payments of money in amounts that are just enough to provide a minimum standard of living, where someone can pay for rent, food and other necessities. Eventually the program seeks to cover the costs of education and the medical cover of every subscriber.

2. Provide support for governments by making a market in Dot Money. Thereby the Global Currency Reserve (GCR) also directly makes a market in GCR eligible currencies and helps to sustain their respective values. In addition, using Dot Money, the GCR intends to assist governments by helping to pay some of the largest costs of governments associated with their social and welfare programs by providing the minimum monthly stipend.

3. Help stabilize the global economy and protect against rapid and extreme devaluations of major GCR eligible currencies by providing a hedging mechanism, that acts as an insurance policy to protect the value of major global currencies at all times, and in particular during times of crisis, war and disaster.

4. Help businesses to save money and lower costs by paying for a portion or the entire amount of their sales taxes to their respective governments when Dot Money is used as the medium of exchange. Sales taxes will be paid to tax authorities in any combination of their native currencies or Dot Money at the discretion of the tax authority themselves.

5. Help reduce the necessity for governments to collect income taxes. Dot Money will help governments to substantially lower taxes or completely eliminate the need for certain taxes, including income taxes.

6. Help mitigate, lower or completely eliminate the national debts of any GCR eligible countries in the world, whose governments permit the use of Dot Money within their borders and accept Dot Money as payment themselves.

7. Provide a tool for ordinary people, businesses, institutions and governments to protect the value of the money they've earned. Today people who are interested in protecting the value of the money that they have worked hard to accumulate are encouraged by various profiteers to become speculators by investing into everything from precious metals, to stocks, bonds, commodities and property. The problem is that, becoming an investor into any of these financial products and commodities exposes the investor to risks associated with the fluctuation of the prices in the global market of these products into which they invest, and the base currencies in which their investments are made.

As a result being an investor in a global marketplace, even investing into secured bank bonds, really does very little to help guarantee that funds invested will have the same buying power in relation to the global economy as they did when they were invested, even if the investment is returned in full with interest. Dot Money is not an investment product but Dot Money is a record of exchange of value for value that is supported and sustained by the users of Dot Money. These records of exchange for value can be exchanged themselves and thus their values are sustained by convention of the users of Dot Money.

Dot Money is not designed to replace local native currencies but to help sustain the market in all GCR eligible currencies. Thus, the only risk to people who convert their native currency into Dot Money would occur if everyone stopped using Dot Money. The exclusive business of the Global Currency Reserve (GCR) is to be the primary market maker and administrator for Dot Money. Other secondary markets for trading and market making in Dot Money in the retail markets also exist.

Thus purchasing Dot Money records (buying Dot Money) provides protection against devaluation that can result from the unregulated fluctuations in the rates of exchange between all global currencies. In other words, people who transact in Dot Money sustain the value of their local currencies at the time of purchase of Dot Money. Their liquid assets that are held in Dot Money records of exchange (the Dot Money "virtual currency") and the Dot Money can be converted at a later date back into any GCR eligible currency at the same fixed rate as when the Dot Money was purchased. Thus Dot Money retains its value in relation to all of the GCR eligible currencies over time. Since Dot Money is accepted around the world it can be spent just like any other money, and there is no need to ever convert from Dot Money directly to any other currency.

8. Implement the Dot Master human based system. The GCR will provide a human based, global financial transaction and clearing method for Dot Money that is not dependant on electronics, the internet or telecommunications. The Dot Money human based system makes use of people called Dot Masters to keep manual copies of the same records held in the Dot Money distributed software network. The Dot Money human based system can be used all over the world including remote locations to buy and sell goods and services, and record exchanges of value using Dot Money. GCR will provide a human based, concurrent accounting and transaction system along with the electronic record keeping system in order to help detect and further reduce fraud, and ensure that the value and utility of Dot Money survives and persists for its users in spite of any manmade or natural disasters or failures of internet technology.

9. Continue to implement methods to receive, hold and spend Dot Money that can be used around the world with or without computers or cell phones and in remote locations. The Dot Money transactions network backbone is being designed to be able to store records and be able to be used without computers, and without physical cards or other physical representations of money in remote locations. This is in order to reduce fraud and reduce vulnerabilities associated with any dependencies on global electronics, communications and internet systems. Dot Money is implanting technology today that will make Dot Money and the Global Currency Reserve (GCR) exist, using a powerful simultaneous backup method consisting of a vast human network of Dot Masters that also keeps the same records that exist in the Dot Money global electronic network. In this way use of Dot Money will be able to take place in areas where no technology exists and all users of Dot Money around the world will have the value of their Dot Money protected from technology failures.

10. Help facilitate all means of global barter exchanges. Dot Money will seek to develop and implement a mechanism whereby the Global Currency Reserve (GCR) can help facilitate international barter trade of goods and services by acting as an intermediary to the trade. Because Dot Money is an international organization and Dot Money is more than just a virtual or abstract currency, the units of Dot Money can be expressed in the form of any amounts of goods or services. In this way a person can just as easily trade cows for value (or any other item or service) instead of Dot Money, but utilizing the Dot Money infrastructure and technology.

Engaging in barter trade is a human practice of establishing and agreeing on value that will likely endure to facilitate trade; that forms the basis of Dot Money. Dot Money and the GCR would like to be part of helping to facilitate and evolve barter practices around the world in order to help the global economy to evolve and realize its full potential.

What can I do with Dot Money?

You can use Dot Money to purchase goods and services around the world in a manner similar to Bit Coin, but with more advanced features that enable transactions with minimal technical overhead and experience. You can convert your Dot Money back into any other Global Currency Reserve (GCR) eligible currency at any time online or in person at any Dot Money exchange or GCR partner bank in the world. However, in order for Dot Money to be as successful as possible, we incentivize everyone to keep their money in Dot Money format as often and as long as possible. In order to further incentivize the holding and transacting in Dot Money, part of the plans of the GCR is to pay the sales taxes of selected goods and services in transactions that are conducted in Dot Money. Since Dot

Money can be used just like money there is no reason to convert it back to other currencies, other than for the purposes of lending, investing and other businesses conducted by banks and financial institutions.

Ultimately, if enough people use Dot Money the Dot Money system will enable subscribers to receive a minimum monthly subsistence payment to use, however they like, in order to help overcome poverty. Current plans are to reach a net amount equivalent to $1,600 USD per month after taxes at the valuation of December 1, 2014. Obviously this facility will be of great assistance to governments whose currency are GCR eligible by helping them to lower the costs associated with their social and welfare practices. **Dot Money is a private sector solution to social welfare that does not require the redistribution of wealth.** Dot Money helps each and every individual in the world as well as their respective governments to unlock their full potential. Those who apply and are approved to receive the monthly stipend will also be allowed to express their desires, concerns and ideas to the administrators of Dot Money through a real time voting system, facilitated by the GCR for use by those who hold Dot Money, at the time that voting is scheduled to take place.

Dot Money can be used by businesses to lower their cost of transactions and the plan is for the Dot Money administrators to eventually pay the sales tax associated with transactions paid for using Dot Money. In this way the costs of businesses will be lower and users of Dot Money will be incentivized to purchase goods and services from vendors that accept Dot Money.

Dot Money can be used by individuals, speculators, businesses, intuitions and governments as a tool that provides a hedge against the devaluation of the major global currencies, specifically those GCR currencies that are eligible for trading in Dot Money.

Dot Money can be a source of business for banks and other financial institutions that facilitate any kind of financial transactions. There are extraordinary opportunities and positive benefits for banks who work directly with Dot Money and the Global Currency Reserve (GCR).

In addition there are lucrative opportunities for individuals, businesses, banks and other organizations who can apply to become Liquidity Providers or work as Dot Masters for the Global Currency Reserve (GCR) and make money by buying and selling Dot Money directly on behalf of the GCR, which is the administrative body of Dot Money.

How do I get some Dot Money for myself?

People who want to obtain Dot Money must register using an email address and utilize the methods listed below to obtain Dot Money for their own account:

1. Invitation. Accept an invitation by the Global Currency Reserve (GCR) to participate in the implementation and administration of Dot Money. Accept an invitation to become a Dot Money Agency (Dot Money authorities for each country) and Dot Master (human based Dot Money network facilitators) who are invited and selected respectively by the management of Dot Money and the GCR to work directly with, and on the behalf of, the GCR in order to administer, promote and facilitate the use of Dot Money in their respective territories. Those who help by participating in the crowdfunding of the Dot Money Project before the official launch of Dot Money will be given preferential consideration for invitation to work with Dot Money in a paid capacity, as well as other potential benefits and rewards. For more information on how you can become a crowdfunder of Dot Money please visit www.DotMoney.Cash.

2. Directly Purchase Dot Money. Purchase Dot Money

directly from the Dot Money administrative body or at participating banks and Dot Money exchanges at the rates of exchange, established by the Dot Money administrative board, after the launch of Dot Money. Dot Money will be offered for sale by the Global Currency Reserve (GCR) after the official launch of Dot Money. For the tentative launch date of Dot Money please go to the GCR website www.DotMoney.Cash.

3. Become a Liquidity Provider. Become a Dot Money "Liquidity Provider" by buying Time Released Dot Money after the launch of Dot Money at a discount rate offered by the Global Currency Reserve (GCR). Time Released (or Time Coded) Dot Money can be only be spent in the future according to the Time Release code on the Dot Money that is delivered to the buyer immediately upon purchase. Liquidity Providers not only help the GCR to sustain the value of Dot Money but also help to stabilize entire the global economy and the value of all eligible GCR currencies (most major currencies plus many others).

4. Accept Dot Money as payment. After the launch of Dot Money you can sell goods and services in exchange for Dot Money. Businesses who want to accept Dot Money as a form of payment can use the Dot Money payment gateway system being developed privately in association with the Global Currency Reserve (GCR).

5. Use Dot Money as a hedge for currency speculation. After the launch date of Dot Money anyone will be able purchase and/or exchange Dot Money around the world in the same way that BitCoins are currently exchanged. Because the exchange rates of Dot Money and all GCR eligible currencies are fixed within a specific exchange rate, Dot Money creates a new and powerful hedging tool for currency traders around the world. Certain Institutional investors, financial institutions and banks can negotiate directly with the GCR for specific rates of exchange for trading.

6. Monthly living allowance. At some point after the

official launch and further development of Dot Money and associated systems, people will be able to apply to receive a minimum monthly payment of Dot Money (or Basic Minimum Income or 'BMI'). The Dot Money paid will be in an amount sufficient to help provide recipients with enough money to pay for food, rent and additional basic necessities. Eventually the system will seek to help cover some of the costs of basic medical insurance, basic communications, transportation and education. The BMI payment can be used for whatever the recipient likes and the income taxes associated with the receipt of the Dot Money will be paid to the government of each respective recipient by the administrative team of Dot Money (if the government requires that this living stipend be taxed).

How does Dot Money work?

Dot Money works as a result of the cooperation of many different individuals, businesses, financial institutions and banks. The most important component of the success of Dot Money is the will of ordinary people to insist on helping to achieve the goals of Dot Money by using, promoting and transacting in Dot Money as much as possible, and encouraging others to do the same. If there is no demand for the ideals, goals and solutions provided for by Dot Money then no-one will use Dot Money, and Dot Money will cease to exist in a similar way that any currency including the U.S. Dollar would cease to have any use if everyone stopped transacting in it. However, wherever there is a sincere will of people to end poverty, to help improve the lives of the poor without taking anything from the wealthy, and wherever there is a desire to have a stable global economy, keep governments from going bankrupt, lower taxes and sustain the value of currencies in which we all transact, then Dot Money and the users of Dot Money will prosper. Dot Money is driven by human will power, and helps to facilitate what people, businesses, and governments all want the most.

As a business Dot Money and the Global Currency Reserve (GCR) is not the same organization. The Global Currency Reserve (GCR) is a private for profit institution that administers the issuance of Dot Money on behalf of users and subscribers of Dot Money. The GCR is an international business corporation, with separate independent corporate agencies that facilitate the implementation and use of Dot Money in GCR eligible countries on behalf of the GCR. The international structure of Dot Money resembles the structure of the U.S. Federal Reserve where member Dot Money Agencies are privately held but regulated by the GCR, and conduct business on behalf of GCR in their respective territories in order to carry on the business and goals of GCR. Dot Master Member Agencies (also referred to as the Primary Dot Masters of their respective territories) receive some profits from the management of money at the direction of GCR, and from the business conducted in their own territories. Primary Dot Masters also receive subsidies and a portion of an annual dividend paid to them by the GCR that is issued as a small percentage in relation to the amount of money being held by the GCR over the course of the year.

Even though the GCR does have some structural similarities and some similar functions to the U.S. Federal Reserve there are major differences between the functions and purposes of the GCR and the U.S. Federal Reserve.

The first major difference is that the mission of the GCR is global in nature and it does not serve the specific interests of any particular country. It attempts to act in cooperation with all governments and businesses around the world that allow for the use of Dot Money to achieve the specific goals of Dot Money for the benefit of mankind. GCR seeks to cooperate with all civilized governments but management intentionally distributes the infrastructure of GCR and Dot Money around the

world in order to reduce the risks of being forcibly manipulated by any single government. GCR is the friend of peace loving governments and people around the world and operates within the boundaries of the laws of the countries in which GCR conducts business. However, **GCR also has a responsibility to remain as impartial as possible with respect to any single GCR eligible country or group of countries. GCR also has a responsibility to persist in the face of any adverse actions that may arise against the GCR.**

Furthermore, the GCR is not a bank, commercial lending institution, brokerage, investment advisor, nor does it store value for any organization other than itself. When Dot Money is issued there is no storage facility necessary. The GCR is a private corporation with associated corporations existing in all countries where Dot Money is legally able to be used. The GCR provides a list on its website of counties whose currencies are eligible for trading by the GCR, along with the rates at which the GCR exchanges Dot Money. In so doing it directly supports and creates a market for each GCR eligible currency. GCR eligible countries are those whose governments work in cooperation with Dot Money and may also include additional countries who allow Dot Money to be used by its citizens and do not expressly outlaw or prohibit the use of Dot Money or make the use of Dot Money onerous.

Dot Money uses some of the same successful software and methodologies as BitCoin with several major enhancements that distinguish Dot Money in function and purpose from BitCoin. Dot Money is created for different purposes than BitCoin and **the fundamental differences are in creation, attainment and rates of exchange of Dot Money as opposed to BitCoin**.

Brand new BitCoins are created and obtained in a similar fashion to mining for gold, except that people work with their computers in an intensive and expensive mining process. Thus Bit Coin is just another commodity, like gold, except that you cannot hold a BitCoin in your hand. Bit Coin is traded on global markets in a similar fashion as any other commodity with prices that fluctuate according to market forces of supply and demand for BitCoins.

Dot Money is created for a different purpose to Bit Coin and thus **the creation of Dot Money is governed by an administrative body (human beings) also called "Dot Masters" under the direction of the Dot Money Global Currency Reserve (GCR).** Anyone or any organization in the world can apply to become a Dot Master. Dot Masters consist of a collection of the owners and administrators of the system as well as people and organizations from all over the world.

To put it simply you can think of the GCR as a parking place for money from around the world, and the value of the units of Dot Money as the prices at which the GCR defends the values of those global currencies and the value of Dot Money. **The GCR keeps the majority of the money that it receives in exchange for selling units of Dot Money called "Dots". The value of Dot Money is determined by the GCR who trades Dot Money within a tightly fixed range of exchange rates for GCR eligible currencies.** Thus the GCR provides users of Dot Money with a powerful hedge against the devaluation of any global currencies in exchange for simply using Dot Money. Although Dot Money can be used in the same way as money, Dot Money is not actually money but an electronic and hard copy record of exchange of value that is defended by the users of Dot Money, the GCR and its member associations, cooperative governments, banks, businesses and most importantly the peace loving people of the world. The rates of currency exchange between Dot Money and the approved GCR eligible currencies that are accepted into the Dot Money Global Currency Reserve (GCR) is determined and adjusted by the Dot Money administrators at the GCR. The GCR works with the relevant governments, banks and stakeholders to maintain optimal rates of exchange. **Dot Money works with banks and governments to help prevent the devaluation of any of the currencies that are eligible and accepted into the Dot Money Global Currency Reserve (GCR) system.**

The value and success of Dot Money substantially depends on the will of the people and organizations that support the goals of Dot Money by simply using Dot Money and insisting that businesses and governments also accept the use of Dot Money to as great an extent as they are willing. The more people that use Dot Money and avoid converting it back to any other global currencies the better Dot Money will be able to achieve its stated goals for the good of all of the people of the world. In addition, the use of, and the success of, Dot Money will help stabilize and strengthen the value of all the other currencies of the world, which will enable governments to provide better and more cost effectively services to their citizens, and local banks to provide more credit and loans according to their own practices.

Other components and partners that are associated with the success of Dot Money and which are utilized by GCR but are not necessarily owned or directly controlled by the GCR **include, global payment systems, gateway service providers, related technology, banks, fund managers, financial institutions, private and publicly owned currency exchanges and currency exchange service providers**.

All of the relationships between GCR and the general public, other than certain purchases or sales of Dot Money and other currencies for the benefit of the account of GCR, are performed in cooperation with businesses and individuals that are properly licensed to facilitate such services in their own countries. Although no institution can ever completely protect itself from the dark side of human nature that results in fraud, **Dot Money and the GCR does its best to establish and maintain relationships with service providers that** are perceived as having integrity, good faith, and good intentions, and that practice good corporate governance abiding within all applicable laws and regulations.

Why do governments support Dot Money?

Dot Money is not a threat to the global business of free floating currency trading and speculation. The Global Currency Reserve (GCR) uses **Dot Money** as a tool to help make a market in the GCR eligible currencies of the world (most major currencies) to help sustain the value of these currencies for the benefit of all people. The GCR uses Dot Money **to provide liquidity and a hedging facility to governments, banks, financial institutions and individuals.**

In addition, it is **one of the goals of the GCR to pay the sales taxes of business transactions that are conducted in Dot Money.**

Finally, it is the express purpose of the GCR to use Dot Money as a means to pay every peace loving person in the world a minimum monthly stipend to ensure a minimum standard of living and to pay any taxes associated with this living stipend if necessary. Thus **the GCR will help reduce, and in some cases eliminate, the heavy expenses incurred by governments in order to provide general welfare services.**

Why do banks support Dot Money?

Dot Money is not a threat to banks, but a facility for banks to help retain as much of their native currency on deposit as they can, at all times, in order to profit from the lending of that currency. Dot Money is not intended to replace the currencies of sovereign countries but to help maximize the value of these currencies. In addition, the GCR does not engage in debt based creation or lending of funds from which banks derive their primary profits. Thus **Dot Money does not compete with banks.** In every instance where Dot Money is being used there will be need to be an equal amount of GCR eligible currencies on deposit in one or more banks in every country where Dot Money is in use. As such Dot Money is not the enemy of the banks and **Dot Money seeks to have as many banks as possible as partners where people can walk into any local bank to buy or sell Dot Money, using the funds in the bank accounts that are held at the bank for the benefit of the GCR.**

The relationship between the GCR and any retail or commercial bank is incredibly simple. **Banks become partners with the GCR by simply allowing GCR to hold accounts at the bank, in order to accept money from people who want to purchase Dot Money, and to enable the GCR to purchase Dot Money for their own account** from ordinary people. **Acceptance of credit cards or use of complex merchant accounts is not necessary** in order for a bank to do business with GCR and Dot Money. In addition, **banks that** are able to **provide GCR with a simple online banking API** (as used by many commercial accounting packages) will **not be bothered by any of the overheads associated with the management of the account** as the GCR will be able to remotely monitor and transact from the accounts.

There are substantial profits to be gained and no downsides or substantial risks for banks who allow the GCR to conduct business in accounts at their bank. Banks stand to gain immediate profit from lending to GCR because GCR simply keeps its money in the bank and only purchases Dot Money from funds drawn from a credit line that is provided by the bank and secured against the cash deposits in the account of GCR in the native currency. In other words, **GCR simply parks money in its account at the participating bank** (in the native currency of the local bank) **and then only withdraws money by drawing down on a**

credit line that is provided by the bank that is secured by the funds on deposit in GCR's account. In this way GCR pays interest to the bank for keeping money on deposit that is spent by GCR in repurchasing Dot Money. GCR only buys Dot Money using credit lines from its partner banks in order to ensure that GCR keeps as much capital in reserve as possible at all times in order to mitigate a "run" on any particular currency. **It costs less money for GCR to pay interest on money that it uses to buy Dot Money than it would if GCR were to buy back Dot Money in any other currency.** In this way GCR utilizes the debt based and lending businesses of their partner banks around the world and allows those banks to make money doing what they do best; lending. And again transaction costs are virtually eliminated for the banks in dealing with GCR because GCR monitors their own bank accounts, deposits and issues expenditures through the banks software API.

Why would business and service providers accept Dot Money?

Only businesses and service providers that are not in favor of helping to solve the problems of global economic prosperity of the world, and all of the individuals in it, would be opposed to accepting Dot Money for payments.

In order to reduce the risks associated with any organization who wants to accept Dot Money as payment, the GCR is finalizing the creation of a payment system and gateway access that will allow vendors who accept payment in Dot Money where all or any portion of the payment is instantly converted into the vendors native currency as a condition to complete the sales transaction. While **the GCR will always encourage businesses to transact and retain their payments in Dot Money as far as possible**, **GCR understands that** the current economic conditions and risks require that, in order for some vendors to accept Dot Money as payment, the **GCR must allow for the payment to be made with**

immediate effect in the native currencies of the vendors that are accepting Dot Money as payment.

In order to incentivize vendors around the world to accept payments using Dot Money, retain their payments in Dot Money and pay their own expenses in Dot Money, **it is the intention of GCR to eventually pay the sales taxes associated with certain transactions conducted in Dot Money**. These sales taxes will be paid directly to the government authorities in their respective countries who are responsible for collecting these taxes and they will be paid in either Dot Money or in the native currency of the tax authorities at their discretion.

Vendors and businesses are further incentivized to accept, use and transact in Dot Money because, by doing so and supporting the goals of Dot Money, they will inevitably lower their own costs of doing business. This is particularly true not just because of the goals of GCR to pay the sales taxes for Dot Money transactions, but also because of the goal of providing a minimum monthly living allowance to all peace loving people in the world. Any business owners who support the goals of GCR will ultimately lower their own costs associated with acquiring and retaining high quality human resources. If employees are already being paid some money by the GCR then they can be happy working for lower or even minimum wages in other jobs because they will be adding to their already existing income by taking a job. In addition, because **people will not need to take jobs out of necessity but of their own free will they will be more likely to provide a higher quality of service while conducting their jobs, as they will not be so worried about money and making ends meet**. People will be free to choose what jobs they really want instead of being forced to take jobs that they don't like just to survive.

Banks, financial institutions, businesses and individuals who work with currencies or speculate

in currencies will be able to use Dot Money as a tool to create a hedge for their trading against the devaluation of any GCR eligible currency (most major currencies).

Why would individuals accept and use Dot Money?

It is the stated goal of the Global Currency Reserve (GCR) and Dot Money to provide a monthly living stipend to each individual in the world, including the rich and the poor, in order to provide a safety net that will enable people to maintain a minimum standard of living, even if they are unemployed for whatever reasons. This is not a new idea. For more information about this and other related subjects please obtain a copy of the book "Dot Money" by Eric Majors from www.DotMoneyBook.com. The goal is to provide every Dot Money subscriber in every GCR eligible country to be paid a net income (after taxes) of **$1,600 USD per month, for the duration of their lives** ($1,600 USD according to the **buying power of the USD as of December 1, 2014**).

The business model of the Global Currency Reserve (GCR) and Dot Money can and will enable this to become a reality only if people take the steps of transacting in Dot Money as much as possible, as often as possible, request that others do the same, and that their respective governments cooperate with the management of the GCR to the maximum extent possible.

The implications of the achievement of the goal of providing monthly stipends and achieving any of the other goals of the GCR are incredibly positive for the entire world. **Imagine some of the implications including not having to worry about having money to pay for food, rent, education, basic medical expenses, etc.** Imagine the reductions in crime that will take place because few people will need to steal in order to survive. Imagine being able to take lower paying jobs without having to feel embarrassed because you like these jobs and without having to worry about how to make ends meet. Imagine being able to take time off to spend with your family if that is your priority or to work a part time job instead of a full time job. **The list of positive results that will impact each peaceful person who receives this monthly living allowance, to do with as they please, goes on and on.** People could use the money to sustain themselves while they start new businesses, or make new inventions, or create new works of art. For more information about the potential positive and negative results of the monthly living stipend, read the book "Dot Money" at www.DotMoneyBook.com.

Throughout the discussions and materials presented by the GCR and Dot Money there is reference to "peace loving people." Because the GCR and Dot Money are private institutions there will be basic minimum requirements and conditions for people to be eligible to receive a monthly living stipend. Using the money for peaceful purposes will be one such condition.

Whether or not the business of GCR leads to the payment of a monthly living stipend for all peaceful subscribers of Dot Money in the world, the initial business of **Dot Money is designed specifically to create a market in the GCR eligible currencies (most major currencies) that will help to protect countries, governments and their citizens from devaluations of their currencies**. This goal provides critical protection to all civilized people of the world from man-made and natural disasters, or economic failures, that are of great concern to most people in the world today. If you support global economic stability, then support and use Dot Money (**for more specific information about the threats to the value of global currencies and the solution of Dot Money, please read the book "Dot Money"**

www.DotMoneyBook.com).

Finally, **the GCR will be able to act as a major liquidity provider for governments and financial institutions that will help to reduce the need for governments to fund their operations through collection of income taxes**. It is very possible that the use of Dot Money and the success of GCR may potentially eliminate the need for income taxes altogether.

GCR and Dot Money is not a tool for the redistribution of wealth. The success of Dot Money and the goals of GCR does not require anything to be taken from anyone. The rich are not asked for anything other than to promote and use Dot Money, and anyone who becomes poor for any reason will have a safety net that will enable them to survive. Business owners will have a steady flow of customers whether or not any jobs are available because everyone will still have their minimum living allowance. Thus the entire global economy will be sustained even under the most difficult conditions.

Dot Money is being designed to be able to survive a complete failure of computer networks and electronics systems. The design of Dot Money will enable transactions in Dot Money in even the most remote areas of the world, where no computers or technology exist at all.

Dot Money provides a means for ordinary people to retain the value of the money that they work so hard to attain. Today people are being encouraged to become speculators in gold, silver, precious metals, property and other commodities, stocks, bonds, CDs or other investment products in order to ensure that their money does not lose value. However, currently **there are no conventional investment products or commodities that can be purchased by ordinary people to protect against the devaluation of global currencies that do not expose the purchaser to the**

risks of speculation. **Dot Money provides something that can be purchased today that will retain its value in relation to CGR eligible currencies (most major currencies) no matter what economic conditions occur**.

In addition, GCR provides the ability for anyone or any organization to become a "Dot Money Liquidity Provider" by buying "Time Release" Dot Money today at discounts specified by GCR (more discounts for longer holding terms). **Imagine being able to buy a type of money, that will not depreciate against the major global currencies, at a discount today in exchange for simply holding the money for a specified period of time before spending it**. Purchasing Time Release (or Time Coded) Dot Money also protects the purchaser from any risks associated with the collapse of any financial institutions as neither **Dot Money** nor Time Coded Dot Money is required to be held in any financial institution but exists simply as a record of value exchanged for value in the GCR global computer networks, which are **backed up by a human network of Dot Masters** that **persist even if the internet and all associated electronic systems in the globe fail**. For more information about the human network of "Dot Masters" please read the book "Dot Money" available at www.DotMoneyBook.com.

Dot Money and the Global Currency Reserve (GCR) are not yet selling Dot Money or Time Coded Dot Money. All forms of Dot Money will only be available to be purchased using GCR eligible currencies after the launch date of Dot Money. The tentative launch date of Dot Money can be found on the Dot Money website (www.DotMoney.Cash).

The opportunity is now for you to do something, in your personal capacity, to help solve the major problems of the world and promote economic and financial stability for you, your loved ones and future generations. Help support the Global Currency Reserve (GCR) by donating today to the

Dot Money Project at www.DotMoney.Cash, obtaining a copy of the book, "Dot Money" by Eric Majors (www.DotMoneyBook.Com and encouraging others to do the same. Making a donation through crowdfunding, buying the book. Sales of this document will help support the implementation of Dot Money.

If you would like to help launch Dot Money you may help crowdfund Dot Money by going to www.DotMoney.Cash.

What Dot Money is not.

Dot Money is not a means to "hide money" or financial transactions from law enforcement. Although the initial implementation of Dot Money uses transaction styles similar to BitCoin, Dot Money is administered by the Global Currency Reserve (GCR), which works in association with governments and their respective law enforcement officers. The GCR helps to support and sustain the integrity of Dot Money, and the governments and the peace loving human beings of the world that are the intended beneficiaries of Dot Money. **The GCR will work with law enforcement whenever possible to help ensure that Dot Money is not used to evade or circumvent any laws**. The administrators of Dot Money, the GCR, work closely with governments, and in support of governments, in order to help create solutions that help to improve the world and not add to problems to the world.

Subscribers to the Basic Minimum Income (BMI) payments must agree to specific terms that will be announced when the time comes to register for the basic monthly income. **One of the terms to receive the monthly living stipend will be that subscribers NOT use their funds for any prohibited transactions that are against the laws of their respective governments or in ways that promote violence**. Recipients of the Basic Minimum Income can be excluded from payments if they do not abide by the terms and conditions of their subscriptions.

Dot Money is not a panacea that will make everyone rich, however, it is hoped that Dot Money will represent the beginning of a new age where people will be free to unlock their individual potential, where money works for people, rather than people just working as slaves to money in order to make ends meet.

More specifically, from a technical and business standpoint, **Dot Money and the GCR are NOT any of the following:**

1. Dot Money is **NOT a bank**, it does not hold accounts for any people other than for the organization itself. Dot Money issues, sells and buys Dot Money for its own account.

2. Dot Money is **NOT an insurance company** but it does have a powerful stabilizing effect on the value of global currencies and does provide protection against devaluation of global currencies.

3. Dot Money is **NOT an investment advisor** and **DOES NOT provide investment advice**.
4. Dot Money is **NOT an Investment Fund** but does work in cooperation with various investment funds and banks.

5. Dot Money is **NOT a stored value card.**

6. Dot Money is **NOT a public company;** it is a privately held corporation.

7. Dot Money is **NOT in the business of making loans and does not compete with lending institutions,** but assist banks in their business of lending in their native currencies.

8. Dot Money is **NOT intended to replace the**

currencies of sovereign nations, but Dot Money helps to sustain the value of global currencies by defending its own value through market making at fixed rates of exchange.

9. Dot Money is **NOT a financial services provider but a private organization** that buys and sells for its own account.

10. Dot Money is **NOT a government issued and regulated currency or money** but a virtual widget that can be bought and sold, and traded for value including currency. At its core Dot Money is an elaborate record and transaction keeping system and database that facilitates global trade for goods and services, and provides financial liquidity for governments, banks and individuals by standing ready to purchase currencies in exchange for records of ownership at specific minimum rates of exchange.

When will the Dot Money system be released and how do I obtain Dot Money?

Dot Money is being rolled out in several Phases on a schedule that can be changed in order to best facilitate the growth of the system**. The initial launch of Dot Money will occur when the Global Currency Reserve (GCR) offers Dot Money and Time Coded Dot Money for sale**. **The launch date can be found on the Dot Money website www.DotMoney.Cash**. The current phases of Dot Money and their tentative times to be achieved are as follows:

Phase 1: Development, Pre-Registration & Crowd Funding.
a. Initial partners, businesses, banks, exchanges and liquidity providers are integrated.
b. Software is built and integrated.
c. Individuals and organizations can pre-register to use Dot Money.

d. Individuals and organizations can help crowd-fund Dot Money.

e. Dot Money Agencies (Dot Masters) and participating banks and merchants are solicited.

f. The Global Currency Reserve (GCR) works with governments to finalize the initial list of countries whose currencies are eligible to be exchanged for Dot Money by the GCR and the specific rates of exchange for each eligible currency. This list of GCR eligible countries that don't restrain the use of Dot Money will be expanded as and when other countries agree to allow Dot Money to be used within their borders.

Phase 2: Release.

a. Dot Money is available to be purchased by subscribers to the system.

b. Any rewards for those who helped to crowd-fund the Dot Money Project will be announced.

c. Dot Money can be used to purchase goods and services from participating vendors and can also be purchased by any currency speculators, traders or funds managers who want to use Dot Money to create hedges in their trading.

d. Dot Money can be converted into other GCR eligible currencies within the rates established by the GCR by buying and selling Dot Money at eligible Dot Money exchanges.

e. Dot Money will work with banks and governments to install identification verification systems at locations around the world that will enable people to sign up to receive the Basic Monthly Income (BMI).

Phase 3: Government Assistance.

a. Administrators of Dot Money will begin to pay the sales taxes to respective governments for any transactions being conducted using Dot Money.

b. Individuals and users of Dot Money from all over the world will be able to register with Dot Money in order to receive a minimum monthly income payment, where Dot Money will also pay any income taxes associated with these payments to the respective governments of each of the individual recipients (if necessary).

Phase 4: Individual Basic Income (BMI) Integrity Tests.

a. Once enough people have signed up for a minimum monthly payment then Dot Money administrators will begin to pay people each month with Dot Money. This can help ensure that nobody in the world needs to fall below a minimum level of existence.

b. The payments will start out at low levels in order to test to make sure that the integrity of the system holds up.

c. Income taxes associated with these monthly BMI payments will be paid by Dot Money to respective governments of Dot Money recipients (if required).

Phase 5: Individual Basic Income is Increased and Adjusted.

a. Based on system integrity tests from Phase 4, minimum monthly payment amounts will be increased each month accordingly until a subsistence level monthly payment is reached.

b. The goal of the Dot Money system is to provide every person in the world with a minimum monthly net payment (after taxes) of at least $1,600 USD based on a valuation of purchasing power as of December 1, 2014, with all associated income taxes paid to the respective governments of the individual recipients.

Phase 6: Expansion of benefits and establish Dot Money supply controls.

a. Eventually direct payments by Dot Money administrators to service providers (insurance companies) or governments will be expanded in order to provide basic medical care for people.

b. Implementation of methodologies to provide for the expansion and contraction of the Dot Money supply, based on the population of the world, will be implemented (i.e. life insurance policies or monthly deductions from supply based on the population of Dot Money subscribers).

For more detailed information about Dot Money please

refer to the book, "Dot Money" by Eric Majors (www.DotMoneyBook.com).

What are the keys to success for Dot Money?

1. *Utility to governments.* Because Dot Money assists governments to provide services similar to social welfare services, which is one of the greatest expenses of government, governments support the use of Dot Money. The Global Currency Reserve (GCR) uses Dot Money to assist governments by providing liquidity for their currencies. This enables governments to reduce their national debts and maintain the value of their native currencies.

2. *Utility to the financial community*. Presently the values of major global currencies are under threat of random, unpredictable and substantial devaluations, which can occur for a variety of reasons including economic, manmade or natural disasters. The Global Currency Reserve (GCR) sets the upper and lower boundaries at which Dot Money will be exchanged for GCR eligible currencies.   Thus Dot Money helps to stabilize the value of all major global currencies and protects countries and their constituents from economic and monetary value collapse, from which they are currently unprotected. Dot Money provides a hedging utility for individuals, institutional traders, exchanges, fund managers and governments that can help protect the value of their own investment portfolio as well as the entire global economy.

3. *Reduce taxes for businesses.* One of the goals of Dot Money is to pay the sales taxes and/or VAT of businesses that accept Dot Money as payment. This should substantially lower costs for all businesses and increase profits.

4. *No Risk Financial Transactions for Vendors and Merchants*.   The GCR financial gateway enables businesses to accept payments in Dot Money where the transactions are completed only after either the instant exchange of Dot Money, or an instant payment to vendors in their native currencies using the Dot Money payment gateway. Venders may also choose to accept instant payments in any combination of Dot Money and their native currencies using the Dot Money payment gateway. There will be major incentives to encourage vendors to accept and hold their payments in the form of Dot Money.

5. *Lower costs for human resources*. When Dot Money begins paying minimum monthly living stipends to subscribers the financial strain associated with paying for human resources and acquiring motivated employees will be relieved.

6. Profits for Banks. Banks stand to profit from Dot Money without the risks associated with other virtual currencies. Banks can make substantial profits by simply allowing the Global Currency Reserve (GCR) to open an account for use by people who want to deposit money to buy Dot Money from GCR. There is no need for banks to provide risky merchant services since Dot Money transacts only in currencies and direct transfers of money. GCR keeps virtually all of the money that it collects in those bank accounts in the native currency of the bank. GCR only makes purchases of Dot Money by using a credit line provided by the bank and secured by the money on deposit at the bank. The GCR purchases Dot Money using credit lines instead of direct cash payments in order to preserve as much capital as possible and to mitigate any potential "runs" on a specific currency.

It is less expensive for the GCR to pay the interest on payments for purchases of Dot Money while a particular currency is in play rather than to pay for purchases of Dot Money directly. Because of the way that the GCR uses its bank accounts and credit lines secured by money on deposit, GCR effectively pays the bank interest for the money that it spends out of its accounts at the bank. In addition, because the GCR and Dot

Money work with governments and law enforcement to help reduce crime, banks need not worry about being bothered with regulatory risks and liabilities that may arise as a result of the sales and purchases of Dot Money. The GCR and Dot Money are also powerful partners for banks to have since they can provide banks with additional liquidity where it is legal and viable to do so.

7. *A new profit stream for currency exchanges*. Brokerages, commodities exchanges and currency exchange businesses will have a new and unique product with which to work. Properly registered, existing currency exchanges and Dot Money currency exchanges will be able to make substantial profits from the spread between the rates at which Dot Money is bought and sold and discounts offered to member exchanges from the Global Currency Reserve (GCR).

8. *Individuals and the will of the people*. In order for Dot Money to be successful (as BitCoin and community currencies, such as the Brixton Pound, are) individual people must take a stand and decide to use and transact in Dot Money as much as possible and to hold their liquid net wealth in the form of Dot Money as far and long as possible, avoiding the conversion back into other currencies wherever possible. Dot Money is not an enemy of any other currency but the evolution of the use of money as a tool, to unlock the true potential of individuals, rather than money that enslaves mankind.

The only thing that creates the value of any money today is the will of the people. Dot Money is not just a means to effectively protect wealth and transact business around the world, but it is a means to potentially end poverty and solve the most pressing problems of the governments, businesses and individuals around the world today.

The goals of Dot Money are to finance a minimum standard of living for everyone in the world without taking anything from anyone, reduce poverty, create global economic stability, stimulate economic growth, increase the quality of goods and services, and lower or end the burden of personal income taxes.

If you support the above stated goals of Dot Money, then do something about it. Help the Global Currency Reserve (GCR) to launch Dot Money as soon as possible. Pre-Register, Sign Up and then use and support Dot Money. Buy the Dot Money book and associated publications at www.DotMoneyBook.Com and www.DotMoney.Cash. Take a stand for yourself and the rest of the world. Help us to unlock your individual potential and to secure the health, happiness and welfare of future generations of people in the world without taking anything from anyone.

9. *The problem solved*. For centuries people have struggled with the economic hardships experienced by those who are poor, including those wealthy people who lose all of their wealth for whatever reason. Social welfare programs founded on taxation and redistribution of wealth are not only unpopular with those who have to pay, but are inadequate due to the limited supply of money in our current global, debt based, monetary system.

Thus, welfare programs that are paid for by the rich and middle class are unsustainable due to the existing debt based monetary supply systems. Use of the debt based monetary supply systems currently in use by governments around the world are now threatening the bankruptcy of these very same governments. Dot Money does not seek to eliminate any of the world currencies or the sovereignty of any of the countries. Rather Dot Money seeks to help provide governments with a "way out" of the problem and a tool that can help them to retain, and in some cases re-establish, the value of their own currencies and the relevancy of their own governments.

None of the current forms of government in the world, upon which we depend, have ever been able to adequately address the problem of the deprivation of rights to use property that arises randomly for people who find themselves in poverty, either through birth or impoverishment for whatever reasons. Every person, both rich and poor must therefore be compensated for the use of property that they would otherwise be able to use to survive (live off the land) if it were not for the existence of the governments, whose primary function is to regulate and protect use and ownership of property. Dot Money will provide this compensation for every people in the world whose governments allow the use of Dot Money within their jurisdictions.

For more information on these and other concepts and why Dot Money will work, please purchase the book, "Dot Money" by Eric Majors at www.DotMoneyBook.com and help crowdfund the launch of Dot Money by visiting www.DotMoney.Cash.

What is the role of Eric Majors in the Dot Money business?

The Dot Money system and implementation was inspired by and is based on the specific idea of "Dot Money", presented in the book "Dot Money", written by Eric Majors. Eric Majors is a promoter, spokesperson and an advisor to the management of the Global Currency Reserve (GCR) and the Dot Money business.

What other individuals, businesses, governments and organizations are partners with Dot Money at the moment?

Dot Money is a large scale global business project and in many instances its successful creation and implementation depends on confidentially between partners and of agreements. Because there are a number of permanent partners and provisional partners, and Dot Money is only accepting Pre-Registration at the moment and is not yet in its released phase, the administrators of Dot Money have left the announcements of many of its key partnerships to be disclosed by the partners themselves if and whenever they choose to disclose their respective relationships with Dot Money. In other cases Dot Money will announce partnerships at such times as management feels such announcements will have an overall positive effect.

The GCR and Dot Money make every effort to coordinate their efforts with the government and law enforcement community responsible for currencies that are eligible for exchange with Dot Money. A specific list of currencies that are eligible to be exchanged for Dot Money by the Global Currency Reserve (GCR) can be found on the website of the GCR (www.GlobalCurrencyReserve.com). GCR eligible currencies that can be used to purchase Dot Money, and into which Dot Money can be converted are those currencies of governments who participate, cooperate with, or support, the Global Currency Reserve (GCR) and the Dot Money program. Dot Money may also be eligible for use with currencies of governments that do not oppose Dot Money and that do not create any regulatory hurdles that make the operation of Dot Money within their country too onerous. **It is important to understand that inclusion on the list of GCR eligible currencies does not necessarily constitute an endorsement of Dot Money or the GCR by any government and does not necessarily constitute a cooperative working relationship or any relationship of any kind between the GCR and the respective governments and their currencies.**

Where is the head office of Dot Money?

Dot Money is a global organization and it is heavily reliant on internet technology. It is intentionally desirous to operate with a globally distributed association of offices at this time. The locations of these offices and any required customer support systems will be announced before the launch of Dot Money. Presently all of the participants and partners in Dot Money have their own offices in different parts of the world and, in some cases, the Global Currency Reserver (GCR) shares offices with Dot Money associates.

The administrators of Dot Money reside in different countries and all activity of the management of Dot Money is conducted and facilitated using the internet. Dot Money (Global Currency Reserve) currently exists as an International Business Corporation formed in the jurisdiction of Belize with associated agencies of Dot Money incorporated in their respective countries of operation to further the business of Dot Money. The primary office addresses where Dot Money receives physical correspondence can be found on the website www.DotMoney.Cash.

How can we help support the launch of Dot Money?

Dot Money is in its initial development Phase and this is a unique time for interested supporters to "get in" on the project. Opportunities to help crowdfund the Dot Money Project can be found on the Dot Money website www.DotMoney.Cash. While Dot Money does already have some substantial, key partners, the administrators and Dot Masters are always open to review and accept offers of help, support and partnerships from anyone who thinks that they have something to contribute. Due to the high number of inquiries that we receive, we simply ask for your patience and indulgence as we may not be able to directly respond to everyone who makes an inquiry.

But if you like the goals of Dot Money and want to really do something to help then here is what we recommend:

1. Crowd Funding. We strongly encourage you to participate in our crowd funding program. By making a (non-tax-deductible) donation to our project you are showing support for the Dot Money ideals and helping to keep the project under the control of ordinary people from around the world. Rest assured that Dot Money will look for ways to reward people who make donations to the cause of Dot Money whenever opportunities arise.

2. Purchase Dot Money products. Dot Money has an online store where it sells the Dot Money book, publications of Dot Money as well as T-shirts and other items. Purchasing any of the items from the Dot Money store helps with the Dot Money Project.

3. Take the Dot Money online polls. From time to time Dot Money creates polls and asks people to cast their votes either for or against certain ideas, including the idea of Dot Money itself. These polls are used by Dot Money in order to help develop the project and to provide politicians from various countries information about the ratio of people that support Dot Money. These polls can be very influential. Just as people pay to cast their votes for their favorite artists on popular televised contests, we believe that it just as important for people to also express their vote on the important world changing business of Dot Money. In addition, charging people some small fee to submit their opinions also helps to ensure that we are getting real votes from real people and not automatically computer generated chaff.

4. Volunteer. Dot Money has an enormous global mission which includes ending poverty. The Dot Money mission involves work in countries all over the world. We welcome help from anyone who believes in what we are doing. If you would like to volunteer to work for Dot Money please go to the website of the GlobalCurrency Reserve.com and click on Partnership Applications and then on Individuals in order to volunteer and to find paying jobs when they are offered by the GCR and Dot Money.

5. New and Events. Please pay attention to the news and events that are being held in relation to Dot Money and the Global Currency Reserve and please sign up and attend these events and stay informed.

6. Make some noise. Register for Dot Money at the website of www.DotMoney.Cash. If you can afford to buy a Dot Money shirt then ware it in public, tell your friends to help support us. From time to time we run awareness campaigns to help inform everyone from politicians, bankers, businesses and ordinary people. Please register for Dot Money and call your representatives in government to encourage them to continue to support Dot Money and the Global Currency Reserve. We may organize some events and ask you to attend in order to show your support for the concepts of Dot Money. Please help support us in any way you can. And please do inform others. Request and send emails to business owners encouraging them to accept Dot Money as soon as it becomes available.

7. Buy and use and hold Dot Money. Obviously, when Dot Money launches and becomes available for purchase, we encourage you to do so.

How can I enter into a partnership with Dot Money or get a job working for the GCR?

Whether you represent a government, bank, currency exchange or a business that wants to accept and transact in Dot Money or you are a person who wants to apply for a job at Dot Money all you need to do is to go to the website of the Global Currency Reserve.com and click on the Partnership Tab and find the correct the correct form on the website to use to communicate with us.

The following tabs are appear on our website for the following people and organizations:

Governments and Large Financial Institutions who are capable of trading $1 Billion US Dollars per year in Dot

Money can gain direct access to the Global Currency Reserve Window and negotiate their rates of exchange directly with the GRC.

Businesses that exchange currencies can apply to become Direct Dealers of Dot Money and buy and sell Dot Money at the Prime Rates directly with the Global Currency Reserve by clicking on the Dealers and Exchanges tab.

High net worth individuals or existing businesses that have enough infrastructure and financial means to support themselves and some new business can apply to become an Agency of the Global Currency Reserve and directly represent the Global Currency Exchange exclusively in specific territories throughout the world and share directly in the profits of the Global Currency Reserve by applying using the GCR Agency tab.

For businesses who would like to accept Dot Money or provide other merchant services in relation to Dot Money there I the Merchants Tab.

Individuals who would like to purchase Dot Money or support the goals of Dot Money by volunteering can do so by selecting the Individuals tab.

Those people who may be interested in applying for a job at the Global Currency Reserve, there is the Jobs tab.

For media outlets who would like to be provided with the press representatives of the Global Currency Exchange there is the Media Inquiries tab .

For Regulators and Law Enforcement agencies who would like to establish a relationship with the Global Currency Reserve or to gain assistance for an investigation into a financial crime there is the Regulators and Law Enforcement Tab.

For all other potential partnership arrangements there is

the others tab.

Another very good way to establish a relationship with the Global Currency Reserve and Dot Money is to attend the events held by the GCR by registering for Dot Money and paying attention to the news and events or by checking the news and events online.

Explain GCR Dot Money Exchange Rates?

When you go to the Global Currency Reserve, the "GCR" web site at **www.GlobalCurrencyReserve.com** you'll see a list of exchange rates between Dot Money and the rest of the currencies in the world.

The first column shows a list of currencies.

The second column shows the TYPE in relation to trading status with the GCR. The TYPE "Primary" indicates that these are currencies that the GRC trades with Dot Money. Currencies labeled Pending are pending acceptance for trading by the GCR.

The next column is the BASE RATE. The BASE RATE is the median price of exchange between Dot Money and all of the other currencies of the world. For example 1 Dot can purchase 1 U.S. Dollar, 1 Dot can purchase 1.23 Australian Dollars, 1 Dot can purchase 0.64 British Pounds and so on.

The next column or the BID RETAIL is the rate of exchange at which the GCR will buy back Dot Money for the currency in the first column from the general public if someone were to purchase Dot Money directly from the Global Currency Reserve.

The next column or the BID PRIM is the rate at which the Global Currency Reserve will buy currency from GCR Direct Dealers. GRC Direct Dealers consist of banks and other retail outlets and online exchanges that buy and sell Dot Money directly from the Global Currency Reserve and with the general public and other businesses. GRC Direct Dealers exchange Dot Money for other currencies as a business for their own account. They make profit by buying Dots at a Discount from the GRC and selling them for more than they purchased them. GRC Direct Dealers also make money by buying back Dot Money for a lower price than what they sell Dot Money.

Qualifying businesses become GRC Direct Dealers by going through an application process on the Global Currency Reserve website and clicking on "Partnership Applications" and then the "Dealers Exchangers" tab.

The next column is the BID INST for Bid Institutional. The BID INST is the average rate at which the GRC buys Dot Money back from Governments and Financial Institutions who are capable of trading over 1 Billion US Dollars per year. Institutional traders of Dot Money have direct access to the GCR window and may negotiate their rates of exchange directly with the GCR for large block trades.

Qualifying Governments, Parastatals and Financial Institutions establish a relationship by going to the GCR website and clicking on the link that says, "Governments, Banks and Institutions."

The next two columns are the BIT STAT or Bid Status and ASK STAT or Ask Status. Under certain circumstances the GCR may suspends buying or selling of any particular currency and if this occurs then an S will appear in either or both the BID STAT or ASK STAT columns. When buying or selling resumes an A for Active will appear in these columns.

ASK INST, ASK PRIME and ASK RETAIL columns represent the prices at which the GCR sells Dot Money to Institutions, Direct Dealers and in the retail markets.

A GCR Eligible Currency is a currency that is eligible for trading by the GCR because the use of Dot Money is permissible and Dot Money is not prohibited or made onerous by the government of the respective country.

Just because a currency is GCR Eligible does not mean that the GCR will engage in trading of Dot Money in that currency and remove the "Pending" TYPE from the currency.

Explain Dot Money Time Release Exchange Rates & Liquidity Providers?

When you go to the Global Currency Reserve, the "GCR" web site you'll see a menu item that says "Time Release Terms." When you click on this menu item you will be taken to a page that shows the cost to purchase Time Release or Time Coded Dots. Buyers of Time Release Dot Money are called "Liquidity Providers."

Liquidity providers help support the GCR by providing capital that enables the GCR to buy Dot Money back and help keep the Dot Money market stable and liquid.

Liquidity providers buy Dot Money today that is time coded so that it can only be spent at a later date. In exchange for this Time Coded or Time Release Dot Money liquidity providers acquire Dots at heavily discounted rates.

Currently the GCR sells Time Release Dots coded with holding terms from 1 year to 10 years. The longer the Term the deeper the discount. Thus it is possible to calculate the annual rate of interest earned when the time coded Dot Money is purchased by the holder.

Because the Time Coded Dot Money is transferred immediately to the purchaser there is no bank or institution that holds the Dot Money on behalf of the purchaser. The record of ownership by the purchaser becomes effective in the Dot Money record system immediately. Thus the purchaser of time Coded Dot Money is protected from the loss of Dot Money that exists independent of the failure of any government or financial institutions.

When you go to the Terms for Time Release Dot page you will see several columns. The first column shows the names of the currencies being traded with Dot Money.

The next column shows the holding TERM in years. Below the value of the TERM is INST ASK. All values below the INST ASK represent the price at which the GCR sells the Time Coded Dot Money to Institutions for the specified TERM in the same column above.

The next column RATE per YEAR shows the Annual Rate of interest that will effectively be "earned" at the end of the TERM when the Dot Money is able to be freely spent.

Below the RATE/YR value there is new title for the values below called PRIME ASK. The values below the PRIME ASK are the prices that the GCR sells Dot Money to GCR Direct Dealers.

The next column, the T RATE is the total discount or cumulative "interest" that would be earned over the total TERM. Below the value of the T RATE is the RETAIL ASK column which is the price that the GRC sells Time Coded Dot Money to the general public when people buy Dot Money directly from the GCR.

All subsequent columns repeat starting with new TERM values.

ABOUT THE AUTHOR

The author, Eric Majors, is an expert in financial market analysis and algorithms, with a Bachelors of Science in Electrical Engineering from the University of Colorado. Mr. Majors is a former U.S. Registered Investment Advisor and business owner who worked with the CIA. He has served as an officer and director of a number of publicly traded companies and as a principal of an international investment banking firm. He is an expert on global currencies and was the principal inventor of Trade Series Management Theory and the associated TSM financial market software systems. Mr. Majors is a certified Consciousness Coach, business advisor, speaker, teacher and author. Mr. Majors is the author of "Financial Markets And Technical Analysis" (2005) and "Dot Money" (2014) and "Dot Money The Global Currency Reserve, Questions and Answers" (2014).

From 2010 to 2013, Mr. Majors spent over 3 years in a U.S. Federal Prison after pleading guilty in 2009 to charges stemming from his work with the CIA and Insider Trading. As a result of the diverse experiences and international exposure that led to his incarceration Mr. Majors is now able to share his unique prospective on life and the global financial markets. He does so from the viewpoint of an insider who speaks candidly and openly.

For more information about please visit:
www.EricMajors.com
www.DotMoneyBook.com
www.DotMoney.Cash
www.GlobalCurrencyReserve.com

ABOUT "DOT MONEY" THE BOOK

Dot Money may be the most important book of our time. It has the potential to transform the world and the lives of every individual for the better. This book explores the creation and use of money, global monetary systems, and our preconceived ideas of money. Then it reveals how ordinary people can take control of the money system today, making it work for them as an alternative to just working to make ends meet. Dot Money is more than a book it is a movement.

Dot Money reveals the next step in the evolution of global economics and shows us how to solve the most important problems of our time. This book has the potential to enable us to overcome poverty, and increase the standard of living for every human being regardless of their current resources, education, race, religion, health, geographic location, political or social affiliations.

For more information please visit:
www.DotMoneyBook.com
www.DotMoney.Cash
www.GlobalCurrencyReserve.com

ABOUT "DOT MONEY, THE GLOBAL CURRENCY RESERVE, QUESTIONS & ANSWERS"

Dot Money is a new and revolutionary kind of global community currency that incorporates the technology of virtual currencies and adds many new features that enable it to be used with or without computers or the internet. The design and purpose of Dot Money is to introduce a new age of economic prosperity and stability throughout the world and solve some of the most important problems facing the world today, including ending poverty. The Global Currency Reserve (GCR) is the international administrator and primary market maker of Dot Money.

The book "Dot Money, The Global Currency Reserve, Questions & Answers" is designed to enable the reader to become familiar with the purposes and functions of Dot Money and the Global Currency Reserve (GCR). This book is designed to be accompanied by the 2014 book "Dot Money" by Eric Majors (www.DotMoneyBook.com). Dot Money may be the most important book of our time.

For more information please visit:
www.DotMoneyBook.com
www.DotMoney.Cash
www.GlobalCurrencyReserve.com